ARCTIC FOXES

BY MARI BOLTE

CREATIVE EDUCATION • CREATIVE PAPERBACKS

Published by Creative Education and Creative Paperbacks
P.O. Box 227, Mankato, Minnesota 56002
Creative Education and Creative Paperbacks
are imprints of The Creative Company
www.thecreativecompany.us

Design by The Design Lab
Art direction by Graham Morgan
Edited by Jill Kalz

Images by Alamy Stock Photo/imageBROKER/Matthias Delle, 21, Sergey Gorshkov, 10; Getty Images/Enrique Aguirre Aves, 9, imageBROKER/Robert Haasmann, 16, John Conrad, 13, Wirestock, 18; Pexels/Irish Heart Photography, cover, 1; Unsplash/Jonatan Pie, 2, 5, 6, Sami Matias, 22–23; Wikimedia Commons/kgleditsch, 12, Lisa Hupp/USFWS, 17, Musicaline, 14

Library of Congress Cataloging-in-Publication Data
Names: Bolte, Mari, author.
Title: Arctic foxes / by Mari Bolte.
Description: Mankato, Minnesota : Creative Education and Creative Paperbacks, [2025] | Series: Amazing animals | Includes bibliographical references and index. | Audience: Ages 6–9 | Audience: Grades 2–3 | Summary: "Discover the made-for-cold Arctic fox! Explore the mammal's anatomy, diet, habitat, and life cycle. Captions, on-page definitions, a Finnish animal folktale, additional resources, and an index support elementary-aged kids"—Provided by publisher.
Identifiers: LCCN 2024010288 (print) | LCCN 2024010289 (ebook) | ISBN 9798889892397 (library binding) | ISBN 9781682776056 (paperback) | ISBN 9798889893509 (ebook)
Subjects: LCSH: Arctic fox—Juvenile literature.
Classification: LCC QL737.C22 B577 2025 (print) | LCC QL737.C22 (ebook) | DDC 599.776/4—dc23/eng/20240405
LC record available at https://lccn.loc.gov/2024010288
LC ebook record available at https://lccn.loc.gov/2024010289

Printed in China

Table of Contents

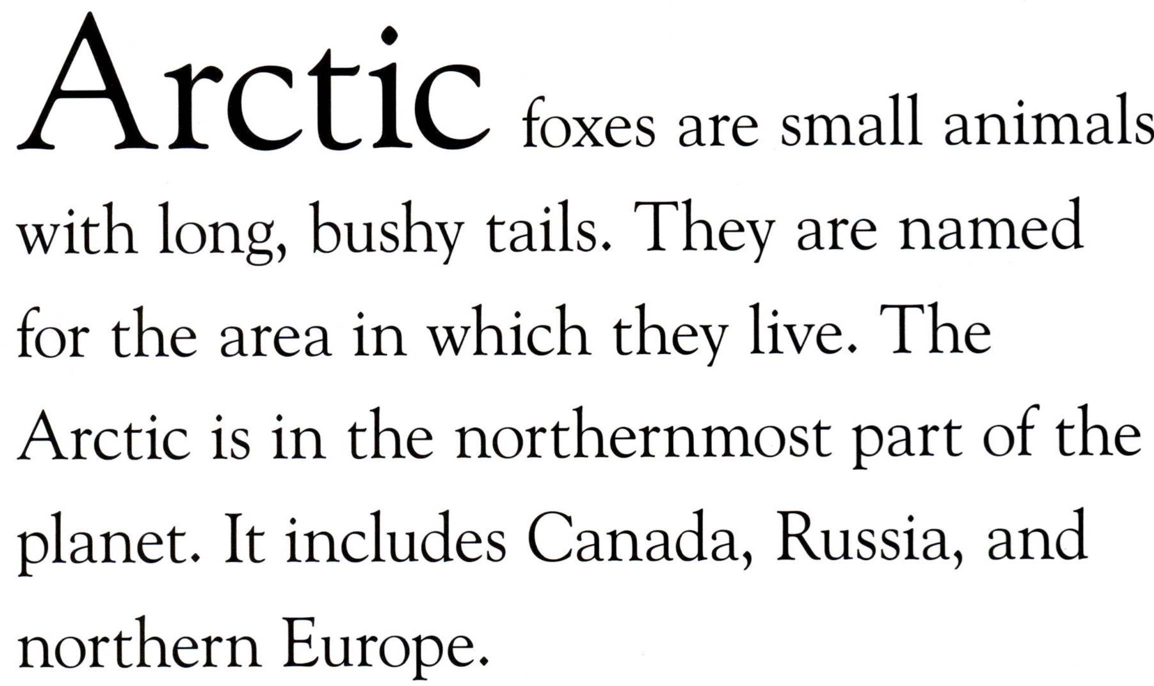

Arctic foxes are small animals with long, bushy tails. They are named for the area in which they live. The Arctic is in the northernmost part of the planet. It includes Canada, Russia, and northern Europe.

An arctic fox's white fur blends in with snow and ice.

Arctic winters are snowy and cold. Summers are cool. Arctic foxes have **adapted** to these conditions. Thick fur coats keep them warm. Short legs, snouts, and ears hold in heat. Furry paws grip the icy ground.

adapted changed to improve the chances of survival

Baby arctic foxes are born with soft, dark brown fur that gets lighter over time.

Arctic foxes are brown or gray in the summer. They match the grass and rocks around them. In the winter, their fur turns white or blue-gray. It blends in with snow and ice. **Camouflage** makes hunting and hiding easier.

camouflage the ability to blend in with the surroundings

Arctic foxes can hear prey moving beneath 4 to 5 inches (10–13 centimeters) of snow.

Arctic foxes weigh between 6 and 10 pounds (2.7–4.5 kilograms). They can walk on top of deep snow. They hear small animals moving below them. With a quick hop, foxes break through the snow and grab their prey.

prey an animal that is killed and eaten by another animal

Lemmings and other mouselike animals are a big part of an arctic fox's diet. So are seabirds. Eggs, berries, and insects are also common foods. If a polar bear leaves part of a seal behind, a fox will grab the easy meal.

Arctic foxes often follow polar bears and eat the bears' food scraps.

*A female arctic fox is
called a vixen.*

Female arctic foxes get ready to give birth in the spring. They build safe spaces called dens. There they deliver five to eight kits. Both parents take care of the kits. Mother foxes feed the kits milk.

kits baby foxes

Kits begin exploring outside the den at three weeks. Their parents teach them how to hunt. Soon after, they start eating meat. Kits are curious. They playfully wrestle and roll around. By six months, they live on their own.

Kits born at the same time to the same mother are called littermates.

Arctic foxes do not stay in one place for long. Young foxes might be forced to leave their family group. Older foxes might be looking for a mate or for food. Arctic foxes usually live three to six years in the wild.

An arctic fox is ready to start a family at about 10 months old.

Arctic foxes may travel thousands of miles looking for their next home. They may travel from one **continent** to another. Scientists track their journeys. It's fun to see how far the foxes go!

continent a large mass of land; Earth has seven continents

An Arctic Fox Tale

An old Finnish story tells of a magical fox. Whoever captured it would be rich and famous beyond their wildest dreams. Every night, the fox ran across the sky. As its bushy tail brushed against the snow, sparks flew. They created dancing waves of light in the sky. These "fox fires" are known today as the Aurora Borealis. They're also called the Northern Lights.

Read More

Gendell, Megan. *Arctic Foxes*. Mendota Heights, Minn.: Apex Editions, 2023.

Marie, Renata. *Arctic Foxes*. Minneapolis: Kaleidoscope, 2022.

Peters, Stephanie. *Polar Bears and Arctic Foxes Team Up!* North Mankato, Minn.: Capstone Press, 2023.

Websites

Britannica Kids: Arctic Fox
https://kids.britannica.com/kids/article/Arctic-fox/627468
Learn more about arctic foxes.

National Geographic Kids: Arctic Foxes
https://kids.nationalgeographic.com/animals/mammals/facts/arctic-fox
Discover cool facts about arctic foxes.

Note: Every effort has been made to ensure that the websites listed above are suitable for children, that they have educational value, and that they contain no inappropriate material. However, because of the nature of the Internet, it is impossible to guarantee that these sites will remain active indefinitely or that their contents will not be altered.

Index